YELLOWSTONE AND GRAND TETON NATIONAL PARKS

UNION PACIFIC CORPORATION

Édition et mise en page : Culturea (Hérault, 34)
Retrouvez notre catalogue sur http://culturea.fr/
Contact : infos@culturea.fr
Imprimé en Allemagne par Books on Demand Gmbh
Design typographique et layout : Derek Murphy et Reedsy
ISBN :9791041985890
Date de parution : février 2024

NATURE'S WONDERLAND

Truly, Yellowstone National Park is a wonderland of nature. Poets have written about it, and artists have painted it, but to believe that such a scenic region is possible, you must see with your own eyes the spouting geysers, bubbling cauldrons, giant canyons and other spectacular phenomena, all of which present a flashing brilliance of motion and color which you will never forget.

Yellowstone is the largest, the oldest and, perhaps, the best known of America's many national parks. It is at once weird, incredible and magnificent in its rugged, wild beauty. It is one of the greatest wild-life sanctuaries in the world.

Frontiersman John Colter discovered the fabulous area by chance in the winter of 1807-08. Colter had been sent by the celebrated trader, Manuel Lisa, to Pierre's Hole in eastern Idaho to make friends with the Crow Indians. Coming home Colter sought a short cut and stumbled upon the mystery of Yellowstone. When he told his friends of the fantastic land he had seen, they refused to believe him.

Twenty years later Jim Bridger brought back a report similar to Colter's, but he, too, was greeted with winks and smiles. Finally, however, the Washburn-Doane Expedition, headed by H. D. Washburn, Surveyor-General of Montana, was organized in August 1870, and accompanied by an army detail, officially investigated the phenomena in the Park.

They encountered innumerable fascinating wonders that even Colter, Bridger and other explorers had missed. They were determined that Yellowstone should be preserved in all its primeval beauty for the enjoyment of all Americans. Their enthusiasm was boundless. Largely because of their efforts a bill was passed through Congress on March 1, 1872, and the rugged wilderness was set aside as Yellowstone National Park.

Since that time millions of travelers have seen the glory that is Yellowstone's. Thousands return year after year. Still others come only once, but they consider it the trip of a lifetime.

GRAND TETON

A few miles south of Yellowstone is Grand Teton National Park, set aside by the government in 1929. Grand Teton and the surrounding country have many distinctive features.

Grand Teton is noted for its mountain grandeur. Many of its jagged, towering peaks of granite rear their heads upward 10 to 13 thousand or more feet. In many respects they resemble the famed Alps of Europe.

Lying at the base of these towering spires are lakes of sapphire-blue, and round about are forests and verdant meadowlands.

Here the visitor feels the tang of the Old West. Central lodges and sleeping cabins are available. Safe saddle trails reach out in all directions.

Motorboats and rowboats as well as trusty saddle ponies may be rented at nominal charges.

SEEING YELLOWSTONE PARK

You may see Yellowstone National Park in your own way. Roam the trails alone, or with an experienced guide, as you prefer. But, by all means, bring your camera!

Yellowstone is your park! It has been provided with fine roads and modern hotels so that you may enjoy its wonders in comfort. You are genuinely welcome to come and stay as long as you like. You could easily stay all summer and never exhaust the alluring appeal, the charm, and tonic effect of Yellowstone's beauty. However, if you can stay for only a limited period, you can still see all the principal points of interest.

You can live as modestly or as lavishly as your choice dictates. In traveling about the Park you may select the style of accommodations that appeals most to you. You can go by Park motor bus with other visitors, or by yourself in a private automobile. Such automobiles are available for hire at government-approved rates.

Splendid camping sites may be found in all parts of the Park. At the main centers of interest are modern hotels, as well as clean, comfortable cabins. Service and accommodations are excellent and one cannot help but marvel at such comfort and luxury in the depths of the wilderness.

So that visitors may see as much of the Park as possible, in a limited amount of time, pre-arranged tours by motor bus are operated. These tours are especially appreciated by folks who travel by rail to Yellowstone. The tour consumes just two and one-half days. Leisure stops are made to properly see all the chief scenic attractions. Another advantage of selecting a pre-arranged tour is that it eliminates the anxiety of driving your own car over unfamiliar mountain roads. Without stress and strain, completely relaxed, you're all the more free to enjoy Yellowstone's wondrous sights.

Bus drivers in Yellowstone Park are thoroughly competent and reliable. Also, Ranger Naturalists of the National Park Service provide many enlightening bits of information on the formations, the flora and fauna in their talks and on guided walks.

WEST YELLOWSTONE ENTRANCE

The Union Pacific route to Yellowstone National Park takes visitors to West Yellowstone, Montana, right on the Park border, and the nearest entrance to Old Faithful Geyser, Yellowstone's most famous attraction. Here, amid the fragrance of the pine forests, the railroad maintains West Yellowstone Station, and an attractive dining lodge.

Upon arriving at the station you have a short time for refreshments; a brief glimpse of the rustic village; mailing cards; buying souvenirs, or perhaps making a few necessary purchases. Then you're ready to climb into a comfortable motor bus for the trip to Old Faithful region.

The first day's ride is a preview of the promise that Yellowstone will fulfill. Along the way you will be constantly amazed by a continuous series of pulse-quickening sights ... green, virgin forests; National Park Mountain; glimpses of the Madison River, and charming views of distant mountains.

Presently the chatter of the cascades of Firehole River is heard. This is one of Yellowstone's most bewitching waterfalls. Oddly enough, while the waters are warmed from geysers and hot springs, the river is alive with trout.

FIRST VIEW OF THE GEYSERS

In the Lower Geyser Basin, Yellowstone parts the curtain and stages a gigantic thermal exhibit. Here the first geysers come into view. There they are! Dipping and skyrocketing; spears of boiling water belch upward, dazzling white against the backdrop of blue skies and forest greenery. For size, number, power and action, no other geysers in the world compare to those of Yellowstone.

In the west and south-central parts of the Park are six major geyser basins. Each contains a number of geysers, pools and springs. This spectacular section of Yellowstone is extremely fascinating. In the Lower Basin are numerous well-known geysers, but in the Upper Basin famous Old Faithful probably will attract your first attention. Moreover, near by is Old Faithful Inn, where you will want to stay for one or more delightful, pleasure-filled days.

Park Ranger Naturalists have prepared an interesting display at Old Faithful which reveals the intricate operation of geysers. An artificial miniature geyser has been constructed and its working parts may be seen and understood. In reality a geyser is a hot spring that has developed into a fountain. A plume of water shoots upward at nearly boiling temperature. The hot volcanic rocks beneath the earth create steam and give the geyser force. At Yellowstone the subterranean action is near the earth's surface and the geysers act as safety valves in draining off this excess energy.

OLD FAITHFUL GEYSER

Most famous of all geysers is Old Faithful. It is everything you would expect to see in a geyser. Old Faithful has the showman's touch, as well. With a rumbling fanfare and roll of drums beneath the thin shell of the earth, Old Faithful goes into action. Then a hissing, boiling pillar of water spurts into the air. Rapidly it gains momentum until it reaches a height of 140 feet or more.

For four minutes it continues to play, and then gradually dies away. Tiny wisps of steam linger a moment, and then disappear. Old Faithful puts on its show within a few minutes of a specified time. The average interval between eruptions is 65 minutes.

It is staggering to conceive just how much water Old Faithful hurls skyward during each of its extraordinary performances. It has been estimated, however, that in a single day this geyser discharges approximately 250,000 gallons of water.

To see Old Faithful at night is, perhaps, the most wondrous sight of all. A giant searchlight is thrown upon its towering plume, and the steaming column of water, thus caught in the vivid, white light, presents a spectacle which becomes an everlasting memory of your trip to Yellowstone.

Numerous other geysers equal or excel Old Faithful in volume or height, but most lack its regularity. The Giant Geyser exceeds all others in the amount of water expelled. Its jet of steaming water sometimes reaches 250 feet, and continues for an entire hour, but it gives no advance notice of when it will burst forth.

Other colorful geysers include the Beehive, Riverside, Grotto, Castle and the Sawmill. Their descriptive names are derived from their fantastic and unusual formations. For sheer individual splendor, the Grand, Giantess and Fountain Geysers are noteworthy.

POOLS AND SPRINGS

When you gaze into Firehole Lake jets of hot gas can readily be seen. Since they look like flame, it is easy to understand why the early trappers called this lake "Firehole."

There are other odd sights in the basins. Fountain Paint Pot looks like a pot of boiling paint, but it is merely colored clay. Morning Glory Pool is so named because of its flower-like cone. Black Sand Pool is a hot spring. Sapphire Springs is accurately named. It is as blue as the gem itself.

After leaving Old Faithful you will soon come to Kepler Cascade. Here a whole series of captivating waterfalls descend about 150 feet, the magic waters singing as they fall.

Continuing on toward Yellowstone Lake—your next stop—you cross the Continental Divide. The Divide crosses the southwest corner of the Park. This immense watershed was created by the tablelands of the Rocky Mountains from which the drainage is easterly or westerly. Eventually the flow on one slope reaches the Atlantic Ocean, and the flow on the other slope finds its way to the Pacific Ocean.

YELLOWSTONE LAKE

Yellowstone Lake covers approximately 139 square miles. Its shore line is 100 miles long. One of the largest mountain lakes in the world, it is a mile and a half above sea level. The blue of its waters against the surrounding snow-capped mountains makes it one of nature's grandest pictures.

FISHING ON YELLOWSTONE LAKE

Countless visitors spend as much time as possible at the Lake. After the first visit this can be readily understood. It is ideal for camping and fishing, and the stimulating air at this altitude is invigorating and delightfully refreshing.

You may take a pleasant boat trip and explore its distant shores or, if you wish, you may engage a motor- or rowboat, at a nominal rate, for a pleasure cruise or to try your luck at angling. Boat trips to Stevenson Island feature fishing, and fish-fries on the Island.

GRAND CANYON OF THE YELLOWSTONE

The Grand Canyon of Yellowstone, as seen from either Artist Point or Inspiration Point, is one of the truly great wonders of the West. It is hard to conceive such breath-taking beauty. Once you have gazed into its jagged depths, alive with color, you will never forget it, nor would you if you could.

Perhaps you stand on the rim, and gaze down into a seemingly bottomless void. Eagles and fish-hawks quite likely will be circling far below. The sides of the ragged pit will be painted with myriads of shifting, changing, vivid colors, with shades of yellow predominating. In the bright sunlight the canyon flames in glory. No sound comes from the echoless, yawning gulf at the bottom.

Grand Canyon—richly named! From Artist Point it is nearly 1,600 feet to the opposite side. There is a sheer drop of nearly 800 feet below the platform on which you stand. Down there lies the green, serpentine Yellowstone River. Your gaze follows the curve of the canyon to where, in the distance, may be seen the silvery sheet of the Lower Falls plummeting downward in a billowy cloud of misty spray.

You can also see the Canyon from other angles and obtain a fuller realization of its majestic beauty. This may be done at Point Lookout, or farther up, at the Grand View.

Time stands still at this mighty chasm. However long you stare in silent wonder, it seems like but a fleeting moment.

THE UPPER AND LOWER FALLS

As the Yellowstone River flows from Yellowstone Lake toward the Missouri River and the Gulf of Mexico, it leisurely twists and winds through the Park until, about fifteen miles from its source, the river is converged to a width of less than fifty feet. Through foaming cataracts it suddenly rushes forward to hurtle down 112 feet in a graceful fall known as the Upper Falls of the Yellowstone. The velocity of flow is so great at the crest that the water pours over the lip of the canyon in a graceful arc.

On approaching the Grand Canyon, a good view of the Upper Falls may be had from an observation point about one quarter mile below Chittenden Bridge.

A short distance beyond the Upper Falls the swift, surging torrent again comes to a mighty precipice, this time with a drop of 308 feet—the famous Lower Falls. Plunging over, it leaps downward with a thunderous roar, and disappears in a cloud of spray, presenting an unforgettable picture. A third of the fall is hidden behind this vast cloud of spray which conceals the mad play of the waters beneath.

Not far from Grand Canyon Hotel is a stairway leading to the top of the Lower Falls. It may also be viewed from the lower end of Uncle Tom's Trail which goes to the bottom of the Canyon. Looking upward at the roaring cataract from that point, you are awed by its wild and unharnessed power. Sunlight gives additional charm to the scene by forming multi-colored rainbows in the filmy clouds of ascending spray.

Leaving Grand Canyon, which you do reluctantly, your route is north to Mammoth Hot Springs. Along the way many interesting sights are seen. About midway between Canyon and Mammoth you see Tower Falls, one of the most graceful waterfalls in the world. Plunging 132 feet into a cavernous basin, rimmed with stately evergreens, the Fall gets its name from the tower-like spires of rock that guard the river's approach to the precipice.

MAMMOTH HOT SPRINGS AREA

From Tower Falls travelers follow the road to Mammoth Hot Springs, site of the Park administration buildings.

Along the sides of the hill, from which the natural springs well, are a group of steps or terraces over which flow the steaming waters of hot springs, laden with minerals. Each descending step has been tinted by the algae (plant life), living in the hot water, in a thousand tones. So vivid are these colors that they appear to vibrate and glow in the sunlight. Some of the older springs have now dried up, but about twenty are still active.

MORE GEYSERS – NORRIS BASIN

Upon leaving Mammoth Hot Springs, on the way back to West Yellowstone, you come upon Norris Geyser Basin. Its geysers spout at frequent intervals and its steam vents noisily erupt great volumes of vapor. Constant, Minute Man and Whirligig are some of the geysers. Emerald, Opal, Iris and Congress are a few of the pools.

Enchanting Gibbon Falls will add another lingering memory of your trip through Norris Geyser Basin.

YELLOWSTONE WILD LIFE

Bears Yellowstone has long been renowned as a refuge for wild animals. While the visitor may not see many animals from the highway, the silent watcher on the trails will not be disappointed.

The famous Yellowstone brown and black bears are the ones most frequently seen. The less-sociable grizzlies are seldom seen.

Park regulations forbid feeding, touching or teasing the bears. Observe them only from a safe distance.

OTHER ANIMALS

Tramping a forest trail your footsteps may disturb a deer that bounds away at your approach. In some of the grassy valleys of Yellowstone are immense herds of elk. Bands of bighorn sheep scale the rocky heights with amazing agility.

Buffalo roam the eastern section of the Park, away from the main roads. Quite likely you will catch glimpses of other animals—antelope, moose, coyote, beaver, porcupine, squirrel and mink, some of which are pictured here.

More than 200 species of birds spend their summers in the Park. Eagles may be seen among the crags. Wild ducks and geese are abundant. Many large, white swans and pelicans lend charm to Yellowstone Lake.

GRASSHOPPER GLACIER

Just outside the northeast corner of the Park is a huge glacier on the surface of which are the broken remains of millions of grasshoppers, preserved through the centuries. Geologists tell us they were trapped here while crossing the mountains in a summer snow storm.

FISHING

Yellowstone is a fisherman's dream come true. Nearly all the streams and lakes contain one or more species of trout. Whitefish and grayling may also be caught. Fishing equipment is obtainable in the Park. No license required.

OTHER FORMS OF RECREATION

Yellowstone offers many forms of healthful recreation. Hiking is popular. Safe horseback trails beckon. Saddle horses and guides are available at Mammoth Hot Springs, Old Faithful and Grand Canyon. Swimming is enjoyed at Old Faithful and Mammoth, where pools, fed by natural warm waters, are maintained. Suits and towels may be rented.

ENTERTAINMENT

Every minute of the day is pleasant at Yellowstone and the evening hours, too, are filled with fun. Impromptu entertainments, lectures, music for dancing and for listening round out the day.

GRAND TETON NATIONAL PARK

Grand Teton National Park became known to white men in 1807-8 when John Colter crossed the range on the memorable trip which resulted in his discovery of Yellowstone. The northern extremity of the Park lies about 11 miles south of Yellowstone's southern boundary.

Grand Teton National Park contains about 96,000 acres, and is penetrated by 90 miles of good trails. Besides its pinnacled peaks and majestic canyons, Grand Teton includes five large lakes and dozens of smaller bodies of water; glaciers, snowfields and a green forest empire of pine, fir and spruce. Much of the Park is above timberline.

The great array of sharp, ragged peaks, which are called the "Teton Range", present some of the grandest mountain scenery in the world. Southwest of Jenny Lake is a cluster of steepled rock, the dominating figure being Grand Teton, the famous mountain after which the Park is named. The towering Grand Teton rises 13,766 feet, 7000 feet above the floor of the valley.

Grand Teton National Park has a rich history. This fertile, green valley and lake region is the historic "Jackson Hole" of pioneer days notoriety, when it was famous as a hideout for outlaws. A large part of it is now included in Jackson Hole National Monument.

The colorful title "Jackson Hole" dates back to 1829 when Capt. Wm. Sublette named it for a fellow trapper, David E. Jackson.

By 1845 the romantic trapper of the "Fur Era" vanished from the Rockies, and during the next four decades the valleys near the Tetons were virtually deserted, except for wandering tribes of Indians who occasionally drifted in. Later government expeditions making surveys of the region named many of the Park's natural beauties—Leigh, Jenny, Taggart, Bradley and Phelps Lakes, and Mount St. John—names which remain today.

The Indian and the outlaw have vanished from this valley but it still retains a flavor of the thrilling Wild West days. The colorful cowboy on spirited pony still rides the range, singing to the cattle, but now he occasionally climbs into the ranch's shiny station wagon and goes over the pass to Victor to meet and bring incoming guests to the ranch, for here are located some of the outstanding "dude ranches" of the West.

For years Jackson Hole has been famous for its big game. In this classification the moose is the most common in the summer. In winter it is the home of the world's largest herd of Wapiti, or American elk. Other wild animals which inhabit the region include bear, mule-deer, elk, Rocky Mountain sheep, beaver, marten, mink, weasel and coyote. Over 100 species of birds have been identified. A unique variety of wild flowers and plants grow profusely in the Tetons. The flowering period begins in the Park as soon as the ridges and flats are free of snow in May and continues until about the middle of August.

Grand Teton National Park is most conveniently reached from Victor, Idaho, on the Union Pacific Railroad. From Victor motor buses climb the Forest Service highway to the top of Teton Pass. This lofty vantage point offers a sublime view of the surrounding domain.

LODGES

Near the town of Moran, Wyoming, is Teton Lodge, and a short distance farther north is Jackson Lake Lodge. Overlooking lovely Jackson Lake, both command marvelous views of the sweeping Teton Range.

The central lodges and cabins are built of native logs. They are equipped with hot and cold running water and are thoroughly comfortable. Trips may be made in all directions from the lodges. Saddle horses, motorboats, rowboats and automobiles may be hired at reasonable rates.

Modern campgrounds for pack trip parties are also available at Jenny and String Lakes. These camps are supplied with running water, sanitary facilities and cooking grates.

DUDE RANCHES

While many of the ranches in the Jackson Hole country are operating cattle ranches, they do accept a few guests during the summer months. All have attractive, comfortable accommodations for those who wish to indulge in horseback riding, fishing, mountain-climbing, hiking and the regular activities of ranch life. Dude ranch life offers rugged outdoor exercise, or pleasant relaxation. Certainly no more ideal spot can be found for such a vacation than in this vicinity. Most of the ranches are located in settings of natural beauty, with mountains, lakes and streams near by. The hospitality of western ranch folk is warm and genuine.

For more complete information about dude ranches in the Union Pacific West, including this area, inquire at any Union Pacific office listed on page 40 of this book for a copy of Union Pacific's book "Dude Ranches Out West".

YELLOWSTONE HOTELS

OLD FAITHFUL GEYSER AND INN

The Yellowstone visitor is well housed and well fed. The hotels furnish modern, comfortable accommodations, and only the finest of foods are served.

OLD FAITHFUL INN

Old Faithful Inn is unique among hotels. Constructed entirely of native logs and stone, it is utterly charming. A massive fireplace in the lounge gives forth warmth and cheerfulness for evening gatherings. Off the lobby is "The Bear Pit," a charming cocktail lounge.

Comparable in appointments to any metropolitan hotel, yet suited to its wilderness setting, is Canyon Hotel near Grand Canyon. Its spacious, glass-enclosed lounge, furnished with comfortable chairs and settees, is a delightful place to relax after a day in the open.

MAMMOTH HOTEL AND COTTAGES

Headquarters of Yellowstone National Park are at Mammoth Hot Springs. Here are located the office of the Park Superintendent and other executive offices of the park administration, and of the public utilities that operate in the park under government regulation and supervision. A post office and museum are located at this point, as is Mammoth Hot Springs Hotel. Mammoth is ideally situated in a charming highland valley among some of the most striking mountain scenery in the Park.

ESCORTED, ALL-EXPENSE TOURS

Why not plan a real carefree vacation this summer by arranging to join one of the congenial groups on an escorted tour, conducted by our Department of Tours? All travel details are taken care of by a courteous, informed escort who accompanies each party. You know in advance exactly what your trip will cost and you are free to enjoy every precious minute of your vacation. The tour parties originate in Chicago and return to that city.

Some of the tours of Yellowstone National Park also include Grand Teton National Park. Others return by way of Colorado, and include the circle tour of Rocky Mountain National Park; others visit the scenic wonderlands of Zion, Bryce Canyon and Grand Canyon National Parks in Southern Utah-Arizona in combination with Yellowstone. There are tours also to California which take in Las Vegas-Hoover Dam, Old Mexico, and Yosemite National Park, as well as tours to the mountain wonderlands of Colorado. There are also tours to the Pacific Northwest, returning through Banff and Lake Louise.

For descriptive literature, reservations, etc., address C&NW-Union Pacific, Department of Tours, 148 So. Clark Street, Chicago 3, Ill., or any Union Pacific or Chicago and North Western representative.

SALT LAKE CITY STOP-OVER

SUNBATHERS BY SALT LAKE

Travelers en route to or from Yellowstone may arrange to stop over at beautiful Salt Lake City. Visitors come from all over the world to see the famous Mormon Temple, or take a dip in Great Salt Lake—a novel experience.

There is a free organ concert daily during the noon-hour for visitors to the Mormon Tabernacle. There are many other interesting sights in and around Salt Lake City.

INDEPENDENT MOTOR BUS TRIPS THROUGH YELLOWSTONE AND GRAND TETON NATIONAL PARKS

Persons who travel independently by railroad make a complete all-expense circle trip of Yellowstone, or in combination with Grand Teton National Park, in comfortable, modern motor buses, operated by competent and informed driver-guides. These circle trips assure your seeing all of the outstanding sights within the Parks. Meals and lodgings are at the hotels in Yellowstone.

SEE SUPPLEMENT FOR COSTS AND SCHEDULES

TICKETS TO YELLOWSTONE PARK VIA UNION PACIFIC

Union Pacific serves West Yellowstone, Montana, directly on the Park's western boundary, and during the Park season operates through sleeping cars from the East direct to West Yellowstone.

During the Park season, also, round trip tickets are sold at nearly all stations in the United States and Canada to West Yellowstone, Montana, or to Victor, Idaho; or, going to West Yellowstone and returning from Victor, or the reverse.

From any place in the United States tickets may be routed so passengers enter the Park at West Yellowstone and depart from Victor, Idaho, Gardiner, Gallatin or Red Lodge, Mont., or Cody, Wyo.—other gateways to the Park.

Traveling Union Pacific to West Yellowstone from the East one can include stopovers at Denver, Ogden and Salt Lake City.

Yellowstone is also a pleasurable side trip from Salt Lake City, Ogden, or Pocatello for travelers to or from the Pacific Coast.

GENERAL INFORMATION

NATIONAL PARK SERVICE—The National Park Service, U. S. Department of the Interior, has full jurisdiction over Yellowstone National Park and is represented by a resident Superintendent, whose headquarters are at Mammoth Hot Springs. The National Park Service, of which The Honorable Newton B. Drury is Director, has jurisdiction over all national parks.

WHAT TO WEAR—Warm clothing should be worn, and one should be prepared for the sudden changes of temperature common at an altitude of 7,500 feet. Visitors should have medium weight overcoats, jackets, "windbreakers" or sweaters. Stout outing shoes are best suited for walking about the geyser formations and terraces, and for mountain use. Women's ordinary street shoes are not well adapted for these walks. Tinted glasses, serviceable gloves and a pair of field or opera glasses will be found useful.

BAGGAGE—The Yellowstone Park Company-Yellowstone Park Lines, Inc., and the Teton Transportation Company will carry free, two pieces of hand baggage for each person, not exceeding 60 lbs. in total weight. Additional pieces of hand baggage, for complete tour of park—$1.00 each, charged by Yellowstone Park Lines, Inc. and by Teton Transportation Co. There is no arrangement for carrying trunks into the Parks.

CHURCH SERVICES—The chapel in Yellowstone National Park is located at Mammoth Hot Springs. Protestant and Catholic services are held every Sunday in the chapel and at other points in the Park, and are bulletined in hotels.

BATH HOUSES—Natural hot-water bathing pools are maintained at Old Faithful and Mammoth Hot Springs. Suit and towel may be rented at a small charge.

MEDICAL FACILITIES—Physicians and a surgeon of long experience have headquarters at Mammoth Hot Springs and are available for service at any place in the Park. Also at Mammoth Hot Springs is a well-equipped hospital with skilled personnel. Trained nurses are also stationed in each hotel. Rates are the same as prevail in cities near the Park.

SADDLE HORSE TRIPS AND GUIDES—Saddle horses and competent guides are available at the Yellowstone Park hotels and Jackson Lake Lodge at reasonable rates approved by the National Park Service. Horseback trips afford opportunities to get far away from roads and beaten paths into the remoter scenic regions and to see many of the more timid wild animals that inhabit the Park. The Dude Ranches in Jackson Hole offer attractive outings.

MAIL, TELEGRAPH AND TELEPHONE—The main post office in the Park is Yellowstone Park, Wyo., and is located at Mammoth Hot Springs. Guests stopping at hotels should have their mail addressed to Old Faithful Store, to Canyon Hotel or to Mammoth Hotel, Yellowstone Park, Wyo., depending at which place the addressee will be when the mail is received. Mail for travelers in Teton Park should be addressed care of Jackson Lake Lodge, Moran, Wyo. Telegraph and telephone service between all hotels; telegraph to all parts of the world; telephone connections throughout the United States. Address your message to the hotel where addressee will be. If the person is at some point other than that of receipt, delivery of message entails a forwarding charge. Money transfers at all hotels in the Park.

SPECIAL AUTOMOBILE SERVICE—The Yellowstone Park Company operates a few sedans for those desiring this special service at additional cost. Advance reservations must be made.

GUIDE AND LECTURE SERVICE—The National Park Service has established a nature guide service at Mammoth Hot Springs, Old Faithful, Lake and Canyon, where guides and lecturers are maintained on the naturalist force to explain and interpret Park features to the public. Trips are made afield, and are so arranged as to be available to everybody. This service is free, as are the evening lectures on the history, geology, flora and fauna.

At Mammoth Hot Springs is a combined museum and information office near the administration headquarters and post office. There are interesting museums also at Madison Junction, Old Faithful, Fishing Bridge and Norris.

CONSULT UNION PACIFIC REPRESENTATIVES—Any Union Pacific representative at the addresses listed on page 40 will be glad to help you plan a trip to Yellowstone Park or to any of the other places reached by the Union Pacific R.R.

FOR THE CAMERA FAN

Yellowstone has everything for the camera addict but the park presents some subjects rather difficult to capture satisfactorily.

In either color or black and white the geysers are best photographed with quartering or slightly back light. Use panchromatic film and a fairly strong yellow filter or even a light red filter.

Color shots of Old Faithful are best made very early in the morning or late in the afternoon. If you are lucky enough to catch a color shot of an eruption against a sunset sky you will have a prize.

Hot pools demand considerable exposure to reproduce the color in the depths of the pool. Do not waste time on hot pools if the day is cool and steam obscures the subject. On such days the geysers give their best photographic eruptions.

In shooting geysers use a fast speed. Expose for the white and let the rest of the picture fall into a low key.

In using an exposure meter in Yellowstone, be careful to see that bright areas in the foreground do not give a false reading.

Your questions regarding either still or moving picture photography in Yellowstone will be answered fully if you will address Manager, Photographic Department, Union Pacific RR, Union Pacific Bldg., 1416 Dodge Street, Omaha 2, Nebr.

HAYNES PICTURE SHOPS—Pictures, albums, guide books, postcards, camera supplies and printing and developing service may be had at Haynes Picture Shops located in the hotels and lodges and elsewhere in the Park.

UNION PACIFIC ·FIRST IN THE WEST

First to link East with West, Union Pacific is still first in furnishing fast, comfortable, dependable, low-cost transportation. During the summer season through sleeping cars operate from Chicago and Salt Lake City to West Yellowstone, making convenient connections, at Ashton, Idaho, for Victor, Idaho, for those desiring to visit Grand Teton National Park only, or in combination with a trip to Yellowstone. For a completely satisfying trip to any of the Western Wonderlands we suggest you Be Specific—Say "Union Pacific".

In addition to Yellowstone, Union Pacific also serves:

A trip to the scenic and magnificent Pacific Northwest can easily be combined with a trip to Yellowstone.

Zion, Bryce Canyon, Grand Canyon National Parks can easily be visited in connection with a trip to Yellowstone, en route to California.

America's foremost vacation and sports center is just a short side trip from Ogden, or Salt Lake City, Utah, or Pocatello, Idaho. Offers a complete summer and winter sports program.

With a variety of climate, scenic attractions and outdoor activities, California provides everything for the perfect vacation. Served by Union Pacific over two routes.

The Rocky Mountain wonderland, where East meets West. Served by fine Union Pacific trains from all parts of the country.

PLAN YOUR TRIP WITH EXPERT HELP

UNION PACFIC
TRAVEL OFFICES

Let one of Union Pacific's courteous and informed representatives assist you with the details of your trip. There is no cost to you and you will find his helpful suggestions will add materially to your enjoyment of the trip. Write, phone, or call at any of the Union Pacific offices listed below.

Aberdeen, Wash. 3 Union Passenger Sta.
Alhambra, Cal. 51 So. Garfield Ave.
Astoria, Ore. 438 Commercial St.
Atlanta 3, Ga. 1432 Healey Bldg.
Bend, Ore. 1054 Bond St.
Beverly Hills, Cal. 9571 Wilshire Blvd.
Birmingham 3, Ala. 701 Brown-Marx Bldg.
Boise, Idaho Idaho Bldg., 212 N. 8th St.
Boston 8, Mass. 294 Washington St.
Bremerton, Wash. 228 First St.
Butte, Mont. 609 Metals Bank Bldg.
Cheyenne, Wyo. 120 W. 16th St.
Chicago 3, Ill. 1 S. LaSalle St.
Cincinnati 2, Ohio 303 Dixie Terminal Bldg.
Cleveland 13, Ohio 1407 Terminal Tower
Dallas 1, Texas 2108 Mercantile Bank Bldg.
Denver 2, Colo. 535 Seventeenth St.
Des Moines 9, Ia. 407 Equitable Bldg.
Detroit 26, Mich. 612 Book Bldg.
East Los Angeles, Cal. 5454 Ferguson Drive
Eugene, Ore. 163 East 12th Ave.
Fresno 1, Cal. 207 Rowell Bldg.
Glendale 3, Cal. 404-1/2 N. Brand Blvd.
Hollywood 28, Cal. 6702 Hollywood Blvd.
Huntington Park, Cal. 7002 Pacific Blvd.
Kansas City 6, Mo. 2 E. Eleventh St.
Las Vegas, Nev. Union Pacific Station

Lewiston, Idaho Room 7, Union Depot
Lincoln 8, Nebr. 234 S. 13th St.
Long Beach 2, Cal. 144 Pine Ave.
Los Angeles 14, Cal. 434 W. Sixth St.
Memphis 3, Tenn. 1137 Sterick Bldg.
Milwaukee 3, Wis. 814 Warner Bldg.
Minneapolis 2, Minn. 890 Northwestern Bank Bldg.
New Orleans 12, La. 210 Baronne St.
New York 20, N. Y., Suite 350 Rockefeller Center, 626 Fifth Ave.
Oakland 12, Cal. 215 Central Bank Bldg.
Ogden, Utah Ben Lomond Hotel Bldg.
Omaha 2, Nebr. Cor. 15th & Dodge Sts. or 1614 Farnam St.
Pasadena 1, Cal. Union Pacific Station
Philadelphia 2, Pa. 904 Girard Trust Bldg.
Pittsburgh 22, Pa. 1419 Oliver Bldg.
Pocatello, Idaho Union Pacific Station
Pomona, Cal. Union Pacific Station
Portland 5, Ore. 701 S. W. Washington St.
Reno, Nev. 209 American Bldg.
Riverside, Cal. Union Pacific Station
St. Joseph 2, Mo. 516 Francis St.
St. Louis 1, Mo. 1223 Ambassador Bldg.
Sacramento 14, Cal. 217 Forum Bldg.
Salt Lake City 1, Utah Hotel Utah, Main and S. Temple Sts.
San Diego 1, Cal. 320 Broadway
San Francisco 2, Cal. Geary at Powell St.
San Jose 13, Cal. 206 First Nat'l Bank Bldg.
San Pedro, Cal. 805 S. Pacific Ave.
Santa Ana, Cal. 305 N. Main St.
Santa Monica, Cal. 307 Santa Monica Blvd.
Seattle 1, Wash. 1300 Fourth Ave.
Spokane 4, Wash. 727 Sprague Ave.
Stockton 6, Cal. 206 California Bldg.
Tacoma 2, Wash. 114 S. Ninth St.
Toronto 1, Ontario 201 Canadian Pacific Bldg.
Tulsa 3, Okla. 823 Kennedy Bldg.
Walla Walla, Wash. First Nat'l Bank Bldg.

Washington 5, D. C. 600 Shoreham Bldg.
Winston-Salem 3, N. C. 632 Reynolds Bldg.
Yakima, Wash. Union Pacific Bldg.
UNION PACIFIC RAILROAD